Contributions of 55 Women Nobel Laureates
Part I - 12 Women Nobel Laureates in Chemistry and Physics

Contents
Preface 4
Chemistry 6
   1.   Marie Curie – 1911 6
      Key Highlights of Marie Curie's 1911 Nobel Prize .. 7
      Term & definition 8
   2.   Irène Joliot-Curie – 1935 10
      Key Highlights of Irène Joliot-Curie's Nobel Prize . 10
      Term & definition 12
   3.   Dorothy Crowfoot Hodgkin – 1964 14
      Key Highlights of Dorothy Crowfoot Hodgkin's Nobel Prize 15
      Term & definition 17
   4.   Ada Yonath – 2009 22
      Key Highlights of Ada Yonath's Nobel Prize 23
      Term & definition 25
   5.   Frances Arnold – 2018 28
      Key Highlights of Frances Arnold's Nobel Prize 29
      Term & definition 31
   6.   Emmanuelle Charpentier – 2020 35
      Key Highlights of Emmanuelle Charpentier's Nobel Prize 36

Contributions of 55 Women Nobel Laureates
Part I - 12 Women Nobel Laureates in Chemistry and Physics

> Term & definition 39
> 7.   Jennifer Doudna – 2020 43
>   Key Highlights of Jennifer Doudna's Nobel Prize .. 44
>   Term & definition (See Emmanuelle Charpentier, section 6, Chemistry) 48
>   8. Carolyn Bertozzi – 2022 49
>   Key Highlights of Carolyn Bertozzi's Nobel Prize .. 50
>   Term & definition 53

Physics... 57
> 1.   Marie Curie – 1903 57
>   Key Highlights of Marie Curie's Nobel Prize 58
>   Term & definition (See Chemistry, section 1, Marie Curie) 60
> 2.   Maria Goeppert Mayer – 1963 61
>   Key Highlights of Maria Goeppert Mayer's Nobel Prize.. 62
>   Term & definition 64
> 3.   Donna Strickland – 2018 67
>   Key Highlights of Donna Strickland's Nobel Prize . 67
>   Term & definition 70
> 4.   Andrea Ghez – 2020 72
>   Key Highlights of Andrea Ghez's Nobel Prize........ 73

Contributions of 55 Women Nobel Laureates
Part I - 12 Women Nobel Laureates in Chemistry and Physics

   Term & definition 75

   5. Anne L'Huillier – 2023 78

      Key Highlights of Anne L'Huillier's Nobel Prize 79

      Term & definition 82

Index 84

Contributions of 55 Women Nobel Laureates
Part I - 12 Women Nobel Laureates in Chemistry and Physics

Preface

Throughout history, women have made Throughout history, women have made the groundbreaking contributions to science, literature, peace, and economics, having risen to the pinnacle of recognition by being awarded the prestigious Nobel Prize. These women Nobel laureates have shattered gender barriers and inspired future generations by excelling in their respective fields. Their work spans diverse areas, from discovering the structure of the ribosome and pioneering ultrafast laser technology to advocating for human rights and global peace.

This "Contributions of 55 Women Nobel Laureates" book explores the profound contributions of these 55 trailblazing women, showcasing their resilience, ingenuity, and dedication to advancing humanity's understanding and well-being. From the first female laureate in 1903 to today, their contributions illustrate the enduring impact of women in shaping the world for a better one.

Contributions of 55 Women Nobel Laureates
Part I - 12 Women Nobel Laureates in Chemistry and Physics

The book describes the key highlights of 55 Nobel Prizes of women and the significance of their works along with the section "Term & Definition" added to the end of each Nobel Prize. This author would like to acknowledge the generous support of "The Nobel Prize" for the photos of women Nobel laureates. "Part I - 12 Women Nobel Laureates in Chemistry and Physics" provides the brief introduction of 12 Nobel Prizes of women in chemistry and physics and the significance of their works.

Thien T. Nguyen, Ph.D.
nguyentrongthien86@gmail.com

Chemistry
1. Marie Curie – 1911

© The Nobel Foundation

Marie Curie's 1911 Nobel Prize in Chemistry was awarded "in recognition of her services to the advancement of chemistry by the discovery of the elements radium and polonium, by the isolation of radium and the study of the nature and compounds of this remarkable element." Curie's contributions not only expanded the scientific understanding of radioactive elements but also had a lasting impact on fields like physics, chemistry, and medicine. Curie remains one of the most iconic figures in the history of science.

Key Highlights of Marie Curie's 1911 Nobel Prize

i. *Discovery of Radium and Polonium*: Marie Curie discovered two new radioactive elements, radium and polonium, which were named after the Latin word for "ray" and Curie's native country, Poland, respectively.

ii. *Isolation of Radium*: Marie Curie's groundbreaking work involved the isolation of pure radium metal, which was one of the most challenging aspects of her research. Curie extracted radium from pitchblende (uranium ore), a process that required processing tons of material to get tiny amounts of radium.

iii. *Advancement of Radioactivity Research*: Curie's work significantly advanced the understanding of radioactivity, a term Curie coined, and demonstrated its potential uses, including medical applications. Curie's research paved the way for using radium in cancer treatments, particularly radiotherapy.

iv. *Second Nobel Prize*: This award made Marie Curie the first person ever to win two Nobel Prizes. Curie had already won the Nobel Prize in Physics in 1903 (shared with Curie's husband, Pierre Curie, and Henri Becquerel) for their combined work on radioactivity.

Term & definition

Cancer— a disease caused by an uncontrolled division of abnormal cells in a part of the body.

Chemistry— the branch of science that deals with the identification of the substances of which matter is composed; the investigation of their properties and the ways in which they interact, combine, and change; and the use of these processes to form new substances.

Medicine— the science or practice of the diagnosis, treatment, and prevention of disease (in technical use often taken to exclude surgery).

Ore— a naturally occurring solid material from which a metal or valuable mineral can be profitably extracted.

Physics— the branch of science concerned with the nature and properties of matter and energy.

Polonium— the chemical element of atomic number 84, a radioactive metal occurring in nature only as a product of radioactive decay of uranium.

Radioactive— emitting or relating to the emission of ionizing radiation or particles.

Radium— the chemical element of atomic number 88, a rare radioactive metal of the alkaline earth series. It was formerly used as a source of radiation for radiotherapy.

2. Irène Joliot-Curie – 1935

Irène Joliot-Curie, the daughter of Marie Curie and Pierre Curie, was awarded the Nobel Prize in Chemistry in 1935 along with her husband, Frédéric Joliot-Curie. They received the prize for their discovery of artificial radioactivity. Irène Joliot-Curie's work, together with Frédéric's, contributed significantly to the development of nuclear science, and their discovery of artificial radioactivity opened new avenues in both basic research and practical applications.

Key Highlights of Irène Joliot-Curie's Nobel Prize

   i. *Artificial Radioactivity*: Irène and Frédéric Joliot-Curie were recognized for creating radioactive isotopes artificially. They discovered that by bombarding certain stable elements with alpha particles (helium nuclei), they could induce radioactivity in those elements. This led to the formation of artificial radioactive isotopes.

ii. *Important Discovery*: Their key experiment involved bombarding aluminum with alpha particles, which produced radioactive phosphorus, marking the first time radioactivity had been artificially induced in a stable element. In addition, they created artificial radioactive nitrogen and silicon in their experiments.

iii. *Impact on Science*: This discovery was groundbreaking because it allowed scientists to create radioactive isotopes of elements that are not naturally radioactive. These artificially created radioactive isotopes had important applications in physics, chemistry, and medicine, including cancer treatment and diagnostic tools.

iv. *Legacy of the Curie Family*: Irène Joliot-Curie's Nobel Prize in Chemistry continued the scientific legacy of the Curie family. Curie's mother, Marie Curie, had won two Nobel Prizes (in Physics and Chemistry), and Irène followed in her footsteps by advancing the field of nuclear physics and chemistry.

Term & definition

Aluminum— the chemical element of atomic number 13, a light silvery-gray metal.

Helium— the chemical element of atomic number 2, an inert gas which is the lightest member of the noble gas series.

Isotope— each of two or more forms of the same element that contain equal numbers of protons but different numbers of neutrons in their nuclei, and hence differ in relative atomic mass but not in chemical properties; in particular, a radioactive form of an element.

Nitrogen— the chemical element of atomic number 7, a colorless, odorless unreactive gas that forms about 78 percent of the earth's atmosphere; liquid nitrogen (made by distilling liquid air) boils at 77.4 kelvins (−195.8°C) and is used as a coolant.

Nuclear physics— the physics of atomic nuclei and their interactions, especially in the generation of nuclear energy.

Phosphorus— the chemical element of atomic number 15, a poisonous, combustible nonmetal which exists in two common allotropic forms, white phosphorus, a yellowish waxy solid which ignites spontaneously in air and glows in the dark, and red phosphorus, a less reactive form used in making matches.

Silicon— the chemical element of atomic number 14, a nonmetal with semiconducting properties, used in making electronic circuits. Pure silicon exists in a shiny dark gray crystalline form and as an amorphous powder.

3. Dorothy Crowfoot Hodgkin – 1964

Dorothy Crowfoot Hodgkin was awarded the Nobel Prize in Chemistry in 1964 for her outstanding work on the structure of important biochemical substances using X-ray crystallography. Hodgkin's citation from the Nobel Committee was "for her determinations by X-ray techniques of the structures of important biochemical substances." Hodgkin's scientific achievements made her one of the most respected and influential chemists of the 20th century.

Key Highlights of Dorothy Crowfoot Hodgkin's Nobel Prize

i. *X-ray Crystallography Pioneer*: Dorothy Hodgkin was a pioneer in the use of X-ray crystallography to determine the three-dimensional structure of molecules. This technique involves bombarding crystalline samples of substances with X-rays and analyzing the diffraction patterns produced to understand the molecular structure.

ii. *Structure of Penicillin*: One of Hodgkin's key achievements was determining the structure of penicillin in 1946. This was a major breakthrough because penicillin had already become a life-saving antibiotic, and understanding its structure was critical for developing synthetic versions and improving its efficacy.

iii. *Structure of Vitamin B12*: Hodgkin's most celebrated discovery, which contributed significantly to her Nobel Prize, was solving the structure of vitamin B12 in 1956. Vitamin B12 is essential for human health, and Hodgkin's work was a major milestone in the field of biochemistry. The structure of vitamin B12 is complex, and solving it was one of the greatest challenges in structural chemistry at the time. Hodgkin's success in this regard provided crucial insights into its biological function.

iv. *Later Work on Insulin*: Although not directly related to her Nobel Prize, Hodgkin's later work focused on the structure of insulin, which Hodgkin pursued for over 30 years. Hodgkin's determination of the structure of insulin in 1969 had profound implications for understanding diabetes and paved the way for the development of insulin treatments.

v. *Contribution to Medicine and Biochemistry*: Hodgkin's research was instrumental in advancing our understanding of biochemical compounds and their functions in the human body. Hodgkin's work has had long-lasting effects in fields such as medicine, pharmacology, and biochemistry.

Significance of Dorothy Crowfoot Hodgkin's Work

i. Dorothy Hodgkin's use of X-ray crystallography revolutionized the study of biological molecules and paved the way for the structural analysis of other complex molecules, such as deoxyribonucleic acid (DNA) and proteins.

ii. Dorothy Hodgkin was the third woman to receive the Nobel Prize in Chemistry, after Marie Curie and Irène Joliot-Curie.

iii. Dorothy Hodgkin's discoveries have had a profound impact on health and medical sciences, influencing the development of drugs and therapies for various diseases.

Term & definition

Antibiotic— a medicine (such as penicillin or its derivatives) that inhibits the growth of or destroys microorganisms.

Biochemical— relating to the chemical processes and substances which occur within living organisms.

Biochemistry— the branch of science concerned with the chemical and physicochemical processes and substances that occur within living organisms.

Biological— relating to living organisms.

Compound— a substance formed from two or more elements chemically united in fixed proportions.

Crystallography— the branch of science concerned with the structure and properties of crystals.

Deoxyribonucleic acid (DNA)— a self-replicating material that is present in nearly all living organisms as the main constituent of chromosomes; it is the carrier of genetic information.

Diabetes— a disease in which the body's ability to produce or respond to the hormone insulin is impaired, resulting in abnormal metabolism of carbohydrates and elevated levels of glucose in the blood and urine.

Diffraction— the process by which a beam of light or other system of waves is spread out as a result of passing through a narrow aperture or across an edge, typically accompanied by interference between the wave forms produced.

Disease— a disorder of structure or function in a human, animal, or plant, especially one that has a known cause and a distinctive group of symptoms, signs, or anatomical changes.

Drug— a medicine or other substance which has a physiological effect when ingested or otherwise introduced into the body.

Efficacy— the ability to produce a desired or intended result.

Insulin— a hormone produced in the pancreas by the islets of Langerhans, which regulates the amount of glucose in the blood. The lack of insulin causes a form of diabetes.

Molecule— a group of atoms bonded together, representing the smallest fundamental unit of a chemical compound that can take part in a chemical reaction.

Penicillin— an antibiotic or group of antibiotics produced naturally by certain blue molds, and now usually prepared synthetically.

Pharmacology— the branch of medicine concerned with the uses, effects, and modes of action of drugs.

Protein— any of a class of nitrogenous organic compounds that have large molecules composed of one or more long chains of amino acids and are an essential part of all living organisms, especially as structural components of body tissues such as muscle, hair, etc., and as enzymes and antibodies.

Structure— the arrangement of and relations between the elements of something complex.

Therapy— treatment intended to relieve or heal a disorder.

Vitamin B12— one of a group of substances (the vitamin B complex) which are essential for the working of certain enzymes in the body and, although not chemically related, are generally found together in the same foods; they include thiamine (vitamin B1), riboflavin (vitamin B2), pyridoxine (vitamin B6), and cyanocobalamin (vitamin B12).

X-ray— an electromagnetic wave of high energy and short wavelength, which is able to pass through many materials opaque to light.

4. Ada Yonath – 2009

Ada Yonath received the Nobel Prize in Chemistry in 2009 for her work on the structure and function of the ribosome. Yonath shared the prize with Venkatraman Ramakrishnan and Thomas A. Steitz. Their citation from the Nobel Committee was "for studies of the structure and function of the ribosome." Yonath's contributions to molecular biology and medicine have had a lasting impact, advancing our understanding of cellular machinery and antibiotic function.

Key Highlights of Ada Yonath's Nobel Prize

i. *Ribosome Research*: Ada Yonath was honored for her pioneering work in determining the structure of the ribosome, the molecular machine in cells responsible for synthesizing proteins. Ribosomes translate genetic information into proteins, which are essential for virtually every function in living organisms.

ii. *X-ray Crystallography*: Yonath used X-ray crystallography to determine the structure of the ribosome at the atomic level. Yonath's work was instrumental in revealing how ribosomes function at a molecular level. The ribosome is a highly complex structure made up of proteins and RNA, and its elucidation was one of the greatest challenges in structural biology.

iii. *Cryo-Crystallography Innovation*: One of Yonath's major contributions to this field was the development of cryo-crystallography techniques, which allowed her to freeze ribosomes in order to study their structure. This was crucial because ribosomes are delicate and prone to damage during crystallographic analysis. The technique involved cooling the ribosome crystals to cryogenic temperatures, enabling their structural study without degradation.

iv. *Insights into Antibiotic Function*: Yonath's work on ribosomes also had significant implications for understanding how antibiotics function. Many antibiotics work by targeting bacterial ribosomes, preventing them from synthesizing proteins. By revealing the structure of the ribosome, Yonath's research provided insights into how antibiotics bind to bacterial ribosomes and inhibit their function, which has helped in the development of new antibiotics.

v. *International Collaboration*: Yonath's groundbreaking research was done through collaborations with scientists worldwide, including her co-laureates Venkatraman Ramakrishnan and Thomas Steitz, who worked independently on similar aspects of ribosomal structure.

Significance of Ada Yonath's Work

i. Understanding the structure of the ribosome has deepened our knowledge of one of the most fundamental processes of life protein synthesis.

ii. Yonath's work is crucial for the development of antibiotics that target bacterial ribosomes, which is especially important in the fight against antibiotic-resistant bacteria.

iii. Ada Yonath became the first woman from the Middle East to win a Nobel Prize in any of the scientific fields and the fourth woman to win the Nobel Prize in Chemistry.

Term & definition

Antibiotic— a medicine (such as penicillin or its derivatives) that inhibits the growth of or destroys microorganisms.

Atomic – relating to the basic unit of a chemical element.

Bacterial – relating to a member of a large group of unicellular microorganisms which have cell walls but lack organelles and an organized nucleus, including some that can cause disease.

Biology— the study of living organisms, divided into many specialized fields that cover their morphology, physiology, anatomy, behavior, origin, and distribution.

Cell— the smallest structural and functional unit of an organism, typically microscopic and consisting of cytoplasm and a nucleus enclosed in a membrane; microscopic organisms typically consist of a single cell, which is either eukaryotic or prokaryotic.

Cellular machinery— the physical and chemical components of the cell that function together to carry out the different physiological functions of the cell.

Cryo-Crystallography— a method used in chemistry to collect diffraction data from crystals at high-intensity X-ray sources by cooling them to form amorphous ice, which helps in preserving the crystal structure for analysis.

Cryogenic— relating to or involving the branch of physics that deals with the production and effects of extremely low temperatures.

Degradation— the condition or process of degrading or being degraded.

Genetic – relating to a distinct sequence of nucleotides forming part of a chromosome, the order of which determines the order of monomers in a polypeptide or nucleic acid molecule which a cell (or virus) may synthesize.

Ribosome— a minute particle consisting of RNA and associated proteins found in large numbers in the cytoplasm of living cells; they bind messenger RNA and transfer RNA to synthesize polypeptides and proteins.

5. Frances Arnold – 2018

Frances Arnold was awarded the Nobel Prize in Chemistry in 2018 for her groundbreaking work in the directed evolution of enzymes. Arnold's pioneering contributions to chemistry and biotechnology have established her as a leader in the field of enzyme evolution and have had a profound impact on sustainable industrial practices.

Key Highlights of Frances Arnold's Nobel Prize

i. *Directed Evolution of Enzymes*: Frances Arnold was awarded the Nobel Prize for developing the technique of directed evolution to create new and improved enzymes. Directed evolution mimics the process of natural selection to evolve proteins or enzymes in the lab with desirable traits. This method allows scientists to accelerate the process of enzyme optimization, which would take much longer in nature.

ii. *First Practical Use of Directed Evolution*: Arnold first demonstrated the directed evolution of enzymes in 1993. Arnold altered enzymes so that they could function in non-natural conditions, such as high temperatures or in organic solvents, which are often needed for industrial processes. The enzymes Arnold developed had improved catalytic properties and could be used in diverse industries, including biofuels, pharmaceuticals, and chemical manufacturing.

iii. *Impact on Sustainable Chemistry*: Arnold's work has had a significant impact on the development of sustainable chemical processes.

The enzymes created through directed evolution can catalyze reactions more efficiently and under milder conditions, reducing the need for harmful chemicals and excessive energy consumption. This makes industrial processes more environmentally friendly and cost-effective.

iv. *Applications in Medicine and Industry*: Directed evolution has numerous applications, including

- *Biofuels*: Arnold's engineered enzymes have been used to develop biofuels, offering a cleaner, renewable energy source.
- *Pharmaceuticals*: Arnold's techniques have also been employed in the production of drugs, particularly in making chemical reactions more efficient and environmentally friendly.
- *Agriculture*: Enzymes created using directed evolution are used in the production of fertilizers and other chemicals important to agriculture.

Recognition as a Trailblazer Frances: Arnold became the fifth woman to win the Nobel Prize in Chemistry, and Arnold's work opened new doors in the field of synthetic biology and protein engineering. Arnold's techniques are now widely used in laboratories and industries worldwide to improve and create new biocatalysts.

Significance of Frances Arnold's Work

i. *Environmentally Friendly Chemistry*: Directed evolution enables the development of enzymes that allow for greener chemical processes, helping to reduce the environmental impact of many industrial reactions.

ii. *Biotechnological Breakthroughs*: Arnold's method has led to numerous advancements in biotechnology, particularly in producing cleaner biofuels and more efficient pharmaceuticals.

iii. *Pioneering Work in Protein Engineering*: Arnold's research has revolutionized protein engineering, offering a powerful tool to design enzymes and proteins for specific tasks, which has broad applications across many fields of science and industry.

Term & definition

Agriculture— the science or practice of farming, including cultivation of the soil for the growing of crops and the rearing of animals to provide food, wool, and other products.

Biocatalyst— a substance, such as an enzyme or hormone, that initiates or increases the rate of a chemical reaction.

Biofuel— a fuel derived directly from living matter.

Biotechnology— the exploitation of biological processes for industrial and other purposes, especially the genetic manipulation of microorganisms for the production of antibiotics, hormones, etc.

Catalytic— relating to or involving the action of a catalyst that increases the rate of a chemical reaction without itself undergoing any permanent chemical change.

Chemistry— the branch of science that deals with the identification of the substances of which matter is composed; the investigation of their properties and the ways in which they interact, combine, and change; and the use of these processes to form new substances.

Enzyme— a substance produced by a living organism which acts as a catalyst to bring about a specific biochemical reaction.

Fertilizer— a chemical or natural substance added to soil or land to increase its productiveness.

Industry— economic activity concerned with the processing of raw materials and manufacture of goods in factories.

Organic solvents— a class of volatile carbon-based chemicals capable of dissolving or dispersing one or more other chemical substances; these chemicals include aliphatic hydrocarbons, halogenated hydrocarbons, aliphatic alcohols, glycols and glycol ethers, and aromatic hydrocarbons.

Pharmaceuticals— a compound manufactured for use as a medicinal drug.

Reaction— a chemical process in which two or more substances act mutually on each other and are changed into different substances, or one substance changes into two or more other substances.

6. Emmanuelle Charpentier – 2020

© Nobel Prize Outreach

Emmanuelle Charpentier was awarded the Nobel Prize in Chemistry in 2020 along with Jennifer Doudna for their development of the clustered regularly interspaced palindromic repeats (CRISPR)-Cas9 gene-editing technology. Their Nobel citation was "for the development of a method for genome editing." Emmanuelle Charpentier's work on CRISPR-Cas9 has had an extraordinary impact on multiple scientific disciplines, making genome editing accessible and enabling new avenues for research and treatment in genetic disorders, agriculture, and beyond.

Key Highlights of Emmanuelle Charpentier's Nobel Prize

i. *CRISPR-Cas9 Gene Editing*: Charpentier and Doudna discovered and developed the CRISPR-Cas9 system, which is a powerful and precise tool for editing deoxyribonucleic acid (DNA). Cas9 is a protein that acts like molecular scissors, capable of cutting strands of DNA at specific locations.

ii. *Discovery of CRISPR-Cas9 Mechanism*: While studying the bacteria Streptococcus pyogenes, Charpentier discovered a previously unknown molecule called tracrRNA, which is involved in the immune defense mechanism of the bacteria. This molecule guides the Cas9 protein to the target DNA. In collaboration with Jennifer Doudna, Charpentier re-engineered this system into a simplified, programmable tool that can cut DNA at any desired location, enabling precise modifications in the genome.

iii. *Revolution in Biotechnology and Medicine*: The CRISPR-Cas9 system has revolutionized the fields of genetics, biotechnology, and medicine by allowing scientists to easily alter DNA sequences in a wide range of organisms. It can be used to:

- Fix genetic mutations that cause diseases such as cystic fibrosis, sickle cell anemia, and muscular dystrophy.
- Study gene function by knocking out or modifying specific genes.
- Develop new therapies for genetic disorders by editing disease-causing genes.
- Improve crops and livestock by enhancing traits like disease resistance and growth.

iv. *Accessibility and Precision*: CRISPR-Cas9 is faster, cheaper, and more accurate than previous genome-editing techniques, making it a widely adopted tool in research labs around the world. It has democratized gene editing, allowing scientists to make specific genetic changes with unparalleled precision.

v. *Ethical Considerations*: The development of CRISPR technology has also sparked discussions about the ethical implications of gene editing, especially in relation to the potential for editing human embryos and the risks of creating "designer babies." Charpentier and Doudna have both called for ethical guidelines to govern the use of this technology, particularly in human applications.

vi. *First All-Female Nobel Chemistry Prize Team*: The 2020 Nobel Prize in Chemistry was notable because it was the first time that an all-female team received the award. Both Charpentier and Doudna are recognized as leading scientists who have made significant contributions to biology and medicine.

Significance of CRISPR-Cas9

i. *Gene Therapy*: CRISPR-Cas9 has opened up new possibilities for treating genetic diseases by editing faulty genes directly in patients' cells.

ii. *Cancer Research*: The technology is being explored in cancer research, where it is used to target and modify genes that drive cancer growth.

iii. *Agriculture*: It has also impacted agriculture by enabling the development of genetically modified crops that are more resilient to pests, diseases, and changing environmental conditions.

iv. *Synthetic Biology*: The ability to edit genomes with precision has implications for synthetic biology, where scientists engineer organisms to produce drugs, biofuels, or other useful compounds.

Term & definition

Cancer— a disease caused by an uncontrolled division of abnormal cells in a part of the body.

Cas9— a 160 kilodalton protein which plays a vital role in the immunological defense of certain bacteria against DNA viruses and plasmids, and is heavily utilized in genetic engineering applications.

Clustered regularly interspaced palindromic repeats (CRISPR)— a family of DNA sequences found in the genomes of prokaryotic organisms such as bacteria and archaea.

CRISPR-Cas9— a laboratory tool used to change or "edit" pieces of a cell's DNA; CRISPR-Cas9 uses a specially designed RNA molecule to guide an enzyme called Cas9 to a specific sequence of DNA; Cas9 then cuts the strands of DNA at that point and removes a small piece, causing a gap in the DNA where a new piece of DNA can be added.

Cystic fibrosis— a genetic condition that affects a protein in the body; people who have cystic fibrosis have a faulty protein that affects the body's cells, its tissues, and the glands that make mucus and sweat; normal mucus is slippery and protects the airways, digestive tract, and other organs and tissues.

Deoxyribonucleic acid (DNA)— a self-replicating material that is present in nearly all living organisms as the main constituent of chromosomes. It is the carrier of genetic information.

Disorder— an illness or condition that disrupts normal physical or mental functions.

Embryo— an unborn or unhatched offspring in the process of development, in particular a human offspring during the period from approximately the second to the eighth week after fertilization (after which it is usually termed a fetus).

Gene— a distinct sequence of nucleotides forming part of a chromosome, the order of which determines the order of monomers in a polypeptide or nucleic acid molecule which a cell (or virus) may synthesize.

Genetic mutation— changes to your DNA sequence that happen during cell division when your cells make copies of themselves.

Genetics— the study of heredity and the variation of inherited characteristics.

Genome— the haploid set of chromosomes in a gamete or microorganism, or in each cell of a multicellular organism.

Muscular dystrophy— a hereditary condition marked by progressive weakening and wasting of the muscles.

Sequence— the order in which amino acid or nucleotide residues are arranged in a protein, DNA, etc.

Sickle cell anemia— one of a group of inherited disorders known as sickle cell disease; it affects the shape of red blood cells, which carry oxygen to all parts of the body; red blood cells are usually round and flexible, so they move easily through blood vessels.

Streptococcus pyogenes— a major human-specific bacterial pathogen that causes a wide array of manifestations ranging from mild localized infections to life-threatening invasive infections.

7. Jennifer Doudna – 2020

© Nobel Prize Outreach

Jennifer Doudna was awarded the Nobel Prize in Chemistry in 2020, along with Emmanuelle Charpentier, for their development of the CRISPR-Cas9 gene-editing technology. The Nobel Committee recognized them "for the development of a method for genome editing." Jennifer Doudna's work on CRISPR-Cas9 has provided scientists with a powerful tool for understanding and manipulating genetic material, with profound implications for science and society. Doudna's research has paved the way for innovations in gene editing, disease treatment, and ethical discussions on genome modification.

Key Highlights of Jennifer Doudna's Nobel Prize

i. *CRISPR-Cas9 Gene Editing*: Doudna, along with Charpentier, was recognized for her role in discovering and developing CRISPR-Cas9, a revolutionary tool that allows scientists to edit DNA with precision, speed, and efficiency. This gene-editing technology has transformed biology and medicine, enabling researchers to alter DNA sequences in virtually any organism.

ii. *Collaboration with Emmanuelle Charpentier*: In 2011, Doudna and Charpentier collaborated to understand the CRISPR-Cas9 system, which was originally a bacterial immune defense mechanism. They re-engineered the bacterial system to create a simplified tool that could be used to cut DNA at targeted sites. This system allows scientists to make precise genetic edits, offering the potential to correct genetic mutations and study gene function in new ways.

iii. *Mechanism of CRISPR-Cas9*: CRISPR is a region of DNA in bacteria that works as a defense system by recognizing and destroying the DNA of viruses. The Cas9 protein acts like molecular scissors, cutting the DNA at precise locations. By designing RNA sequences that guide Cas9 to a specific target, scientists can use the system to edit genes in living cells.

iv. *Revolutionizing Genetic Engineering*: The discovery of CRISPR-Cas9 has had an enormous impact on various fields, including

• *Medicine*: CRISPR is being used to develop therapies for genetic diseases such as cystic fibrosis, sickle cell anemia, and muscular dystrophy. In addition, it holds potential in cancer therapy by targeting specific cancer-related genes.

• *Agriculture*: The technology is used to create crops with improved traits, such as disease resistance, drought tolerance, and enhanced nutritional value.

• *Basic Research*: CRISPR has become an essential tool in laboratories for studying the function of specific genes and understanding complex biological processes.

v. *Ethical and Social Implications*: Doudna has been an advocate for ethical discussions around the use of CRISPR-Cas9, particularly in relation to human genome editing. Doudna has raised concerns about the possibility of gene editing in embryos and the potential for unintended consequences, urging the scientific community to establish guidelines and regulations for its use.

vi. *First All-Female Team to Win Nobel Chemistry Prize*: The 2020 Nobel Prize in Chemistry was historic as it was the first time the prize was awarded to two women without male collaborators. This milestone highlighted the contributions of women in science, particularly in fields like biochemistry and molecular biology.

Significance of CRISPR-Cas9

i. *Gene Therapy*: CRISPR holds immense potential for gene therapy, enabling the correction of genetic mutations that cause hereditary diseases.

ii. *Cancer Treatment*: It is being explored as a tool to enhance immune cells' ability to target and destroy cancer cells.

iii. *Agriculture*: The technology is widely used to develop genetically modified crops with better resistance to diseases, pests, and environmental stresses.

iv. *Ethical Discussions*: CRISPR's ability to edit human DNA has led to discussions about the ethical implications of modifying the human genome, especially concerning potential misuse in creating "designer babies."

Impact of Jennifer Doudna's Work

i. *Global Influence*: Doudna's work has revolutionized the field of molecular biology and continues to shape the future of genetic engineering and biotechnology.

ii. *Innovation and Leadership*: As one of the leaders in the CRISPR revolution, Doudna's contributions have opened new possibilities in medicine, agriculture, and fundamental biology.

Term & definition (See Emmanuelle Charpentier, section 6, Chemistry)

8. Carolyn Bertozzi – 2022

Carolyn Bertozzi, an American chemist, was awarded the Nobel Prize in Chemistry in 2022, sharing the honor with Morten Meldal and K. Barry Sharpless. Bertozzi was recognized for her pioneering work in the development of bioorthogonal chemistry—a field Bertozzi helped establish, which involves chemical reactions that can occur inside living organisms without interfering with the normal biochemistry of the cell. Bertozzi's Nobel Prize marks a significant leap in chemical biology and medicine, laying the groundwork for next-generation therapies, diagnostics, and medical technologies.

Key Highlights of Carolyn Bertozzi's Nobel Prize

i. *Awarded for Bioorthogonal Chemistry*: Carolyn Bertozzi was recognized for pioneering bioorthogonal chemistry, a field she helped establish that enables chemical reactions to take place inside living organisms without disrupting biological processes.

ii. *Impact on Cancer Research*: Bertozzi's work has been instrumental in developing targeted cancer therapies. Bioorthogonal reactions allow drugs to be selectively delivered to cancer cells, minimizing damage to healthy tissues and improving treatment effectiveness.

iii. *Innovative Techniques in Glycobiology*: Bertozzi's research focused on glycans (sugar molecules on cell surfaces), which are key to cellular communication and disease progression. Her techniques allow scientists to track glycans and study their role in various diseases, especially cancer.

    iv. *Collaborative Nobel Prize*: Bertozzi shared the prize with Morten Meldal and K. Barry Sharpless, who contributed to click chemistry, a complementary tool used in chemical biology. Together, their work has revolutionized molecular tracking and drug design.

    v. *First Woman Honored for Bioorthogonal Chemistry*: Bertozzi is the first woman to receive the Nobel Prize for bioorthogonal chemistry, solidifying her position as a leader in the field and an inspiration for women in science.

Significance of Carolyn Bertozzi's Work

    i. *Revolutionizing Chemical Biology*: Bioorthogonal chemistry provides scientists with new ways to study and manipulate biological molecules inside living systems, opening up real-time monitoring of cellular processes and the development of more precise therapeutic interventions.

    ii. *Advancing Targeted Therapies*: Bertozzi's innovations have had a direct impact on personalized medicine, particularly in cancer treatment. By selectively targeting specific molecules or cells, bioorthogonal chemistry reduces side effects and improves the precision of therapies.

iii. *Enhancing Diagnostics and Molecular Imaging*: Bioorthogonal reactions are used to develop advanced diagnostic techniques that allow scientists to label and track molecules, leading to earlier detection and better monitoring of diseases.

iv. *Expanding Applications in Medical*: Research Bertozzi's work has broad applications beyond cancer, influencing fields such as immunology, infectious diseases, and neuroscience. The tools and techniques developed in Bertozzi's lab are now standard in many areas of biological research.

v. *Inspiration for Future Scientists*: As a leading figure in the scientific community and a trailblazer for women in STEM, Bertozzi's success highlights the importance of innovation in interdisciplinary fields like chemistry and biology, encouraging future generations to pursue novel research avenues.

Term & definition

Biochemistry— the branch of science concerned with the chemical and physicochemical processes and substances that occur within living organisms.

Biology— the study of living organisms, divided into many specialized fields that cover their morphology, physiology, anatomy, behavior, origin, and distribution.

Bioorthogonal chemistry— referring to any chemical reaction that can occur inside of living systems without interfering with native biochemical processes.

Cancer— a disease caused by an uncontrolled division of abnormal cells in a part of the body.

Cell— the smallest structural and functional unit of an organism, typically microscopic and consisting of cytoplasm and a nucleus enclosed in a membrane; microscopic organisms typically consist of a single cell, which is either eukaryotic or prokaryotic.

Click chemistry— an approach to chemical synthesis that emphasizes efficiency, simplicity, selectivity, and modularity in chemical processes used to join molecular building blocks; the word "click" referred to easily joining molecular building blocks as two pieces of a seat belt buckle.

Diagnostics— a distinctive symptom or characteristic.

Disease— a disorder of structure or function in a human, animal, or plant, especially one that has a known cause and a distinctive group of symptoms, signs, or anatomical changes.

Drug— a medicine or other substance which has a physiological effect when ingested or otherwise introduced into the body.

Drug design— the inventive process of finding new medications based on the knowledge of a biological target.

Glycobiology— the study of the structure, biosynthesis, biology, and evolution of saccharides (also called carbohydrates, sugar chains, or glycans) that are widely distributed in nature and of the proteins that recognize them.

Immunology— the branch of medicine and biology concerned with the state or quality of being resistant to a particular infectious disease or pathogen.

Infectious— (of a disease or disease-causing organism) likely to be transmitted to people, organisms, etc., through the environment.

Medicine— the science or practice of the diagnosis, treatment, and prevention of disease (in technical use often taken to exclude surgery).

Molecule— a group of atoms bonded together, representing the smallest fundamental unit of a chemical compound that can take part in a chemical reaction.

Neuroscience— any or all of the sciences, such as neurochemistry—the branch of biochemistry concerned with the processes occurring in nerve tissue and the nervous system—and experimental psychology, which deals with the structure or function of the nervous system and brain.

Side effect— a secondary, typically undesirable effect of a drug or medical treatment.

Therapy— treatment intended to relieve or heal a disorder.

Tissue— any of the distinct types of material of which animals or plants are made, consisting of specialized cells and their products.

Contributions of 55 Women Nobel Laureates
Part I - 12 Women Nobel Laureates in Chemistry and Physics

Physics

1. Marie Curie – 1903

© Nobel Prize Outreach AB

Marie Curie was awarded the 1903 Nobel Prize in Physics for her groundbreaking research on radioactivity, which Curie shared with her husband Pierre Curie and Henri Becquerel. Curie's Nobel Prize in Physics honored her profound contributions to science and marked a transformative moment in the history of physics, with implications spanning from nuclear physics to medical treatments.

Key Highlights of Marie Curie's Nobel Prize

i. *Award for Pioneering Work in Radioactivity*: Marie Curie was awarded the Nobel Prize in Physics for their collective work on radioactivity, a phenomenon first discovered by Becquerel in 1896.

ii. *Curie's Contribution*: Marie Curie extended Becquerel's discovery by isolating radioactive elements. Curie coined the term "radioactivity" and demonstrated that radiation was not a result of chemical interactions but generated from the atomic structure itself.

iii. *Groundbreaking Discovery of Radium and Polonium*: In Curie's research, Curie discovered two new radioactive elements: radium and polonium. These discoveries were monumental in advancing the understanding of atomic physics.

iv. *First Woman to Win the Nobel Prize*: Marie Curie became the first woman to win a Nobel Prize, a landmark achievement for women in science, and it established her as a trailblazer for female scientists.

v. *Foundation for Nuclear Physics and Chemistry*: Curie's research laid the groundwork for the future development of nuclear physics, atomic theory, and quantum mechanics, influencing the study of subatomic particles and nuclear energy.

vi. *Personal Sacrifice*: Curie conducted much of her work in physically challenging conditions, often working with hazardous radioactive materials that would eventually lead to her health problems. Curie's dedication set an example of perseverance and passion for scientific discovery.

Significance of Marie Curie's Work

i. *Advancing Atomic Theory*: Curie's work on radioactivity transformed the scientific understanding of atoms, leading to the realization that atoms were not indivisible, as previously thought. Curie's research paved the way for quantum physics and the study of subatomic particles.

ii. *Medical Applications*: The discovery of radium revolutionized medical treatments, particularly in developing radiation therapy for cancer treatment, a legacy that still benefits modern medicine.

iii. *Breaking Gender Barriers*: As the first female Nobel laureate, Curie broke significant barriers for women in science, inspiring generations of females to pursue the scientific careers traditionally dominated by men.

iv. *Dual Nobel Laureate*: Curie's later achievement of winning a second Nobel Prize in 1911 in Chemistry for her work in radioactivity further solidified her as a scientific icon, making her the first and only woman to win Nobel Prizes in two different scientific fields.

Term & definition (See Chemistry, section 1, Marie Curie)

2. Maria Goeppert Mayer – 1963

Maria Goeppert Mayer was awarded the Nobel Prize in Physics in 1963 for her pioneering work on the nuclear shell model of the atomic nucleus. Mayer shared the prize with J. Hans D. Jensen, who independently developed a similar model, and Eugene Wigner, who contributed to nuclear physics. Maria Goeppert Mayer's groundbreaking work on the nuclear shell model earned her a place among the most influential physicists of the 20th century, making significant contributions to our understanding of atomic structure.

Key Highlights of Maria Goeppert Mayer's Nobel Prize

i. *Nuclear Shell Model*: Goeppert Mayer was honored for her development of the nuclear shell model, which explained the arrangement of protons and neutrons within an atomic nucleus. Just as electrons in an atom are organized into shells around the nucleus, Mayer proposed that protons and neutrons are similarly arranged in shells within the nucleus.

ii. *Magic Numbers*: Mayer's most significant contribution was the discovery of so-called "magic numbers", which correspond to numbers of protons or neutrons in the nucleus that lead to particularly stable configurations. These magic numbers are 2, 8, 20, 28, 50, 82, and 126. Nuclei with these numbers of protons or neutrons are more stable than others, similar to the noble gases in the periodic table, which have filled electron shells and are chemically inert.

iii. *Independent Work*: Although Goeppert Mayer collaborated with other physicists, Mayer's breakthrough in the nuclear shell model came independently. Mayer's work provided a theoretical foundation for why certain atomic nuclei are more stable than others, which was a key question in nuclear physics at the time.

iv. *Collaboration with Hans Jensen*: At the same time, J. Hans D. Jensen developed a similar shell model of the nucleus. When they learned about each other's work, they collaborated on a book about the subject, which was published in 1955. Their combined contributions helped explain many experimental findings in nuclear physics.

v. *Only the Second Female Physics Laureate Maria*: Goeppert Mayer was only the second woman to win the Nobel Prize in Physics, after Marie Curie (1903). This achievement highlighted Mayer's remarkable contributions to science, particularly in a field that was dominated by men at the time.

vi. *Legacy in Nuclear Physics*: The nuclear shell model has become a cornerstone of nuclear physics, helping scientists understand the structure of atomic nuclei and explaining the behavior of isotopes, radioactivity, and nuclear reactions. It is still widely used in research today.

Significance of Maria Goeppert Mayer's Work

i. *Stability of Nuclei*: The nuclear shell model explained why certain isotopes are more stable than others, which has applications in nuclear energy and medical technologies.

ii. *Magic Numbers and Nucleosynthesis*: Mayer's discovery of magic numbers helped explain the process of nucleosynthesis, or how elements are formed in stars, particularly in the creation of heavy elements.

iii. *Inspiration for Future Generations Goeppert*: Mayer's achievements inspired many young women to pursue careers in physics and the sciences, overcoming barriers that existed for women in academia at the time.

Term & definition

Atomic nucleus— the small, dense region consisting of protons and neutrons at the center of an atom, discovered in 1911 by Ernest Rutherford.

Atomic structure— the make-up of an atom and what it consists of; an atom is a central positively charged nucleus that is made of protons and neutrons, surrounded by a number of electrons that differs depending on the element of the periodic table.

Configuration— an arrangement of elements in a particular form, figure, or combination.

Electron— a stable subatomic particle with a charge of negative electricity, found in all atoms and acting as the primary carrier of electricity in solids.

Isotope— each of two or more forms of the same element that contain equal numbers of protons but different numbers of neutrons in their nuclei, and hence differ in relative atomic mass but not in chemical properties; in particular, a radioactive form of an element.

Neutron— a subatomic particle of about the same mass as a proton but without an electric charge, present in all atomic nuclei except those of ordinary hydrogen.

Noble gas— any of the gaseous elements helium, neon, argon, krypton, xenon, and radon in the periodic table; they were long believed to be totally unreactive but compounds of xenon, krypton, and radon are now known.

Nuclear physics— the branch of physics that studies atomic nuclei and their constituents and interactions; examples of nuclear interactions or nuclear reactions include radioactive decay, nuclear fusion and fission.

Nuclear reaction— a change in the identity or characteristics of an atomic nucleus that results when it is bombarded with an energetic particle, as in fission, fusion, or radioactive decay.

Nucleosynthesis— the cosmic formation of atoms more complex than the hydrogen atom.

Proton— a stable subatomic particle occurring in all atomic nuclei, with a positive electric charge equal in magnitude to that of an electron, but of opposite sign.

Radioactivity— the emission of ionizing radiation or particles caused by the spontaneous disintegration of atomic nuclei.

3. Donna Strickland – 2018

Donna Strickland was awarded the Nobel Prize in Physics in 2018 along with Arthur Ashkin and Gérard Mourou for their work on laser physics. Donna Strickland's work on ultrafast lasers and Chirped Pulse Amplification (CPA) has made a significant impact across multiple fields, demonstrating the transformative power of advanced laser technologies.

Key Highlights of Donna Strickland's Nobel Prize

i. *Laser Technology*: Donna Strickland was recognized for her contributions to the development of ultrafast laser technology.

Strickland's work focused on the technique called chirped pulse amplification (CPA), which allows for the creation of extremely short and high-intensity laser pulses.

ii. *CPA*: Strickland, along with Gérard Mourou, developed CPA in the late 1980s. CPA involves stretching a laser pulse to lengthen it, amplifying the pulse to increase its energy, and then compressing it back to its original duration. This process produces laser pulses of extraordinary intensity and short duration, which have numerous applications in science, medicine, and industry.

iii. *Impact on Science and Technology*: The CPA technique has had a profound impact on various fields

- *Medical Applications*: CPA lasers are used in precision eye surgeries, such as Laser-Assisted in situ Keratomileusis (LASIK), to correct vision with minimal damage to surrounding tissues.

- *Industrial Applications*: High-intensity laser pulses are used in manufacturing for cutting, welding, and other precision tasks.

- *Fundamental Research*: The ability to generate ultrafast laser pulses has enabled new research in areas such as atomic and molecular physics, providing insights into the behavior of electrons and atoms on extremely short timescales.

iv. *First Woman to Win the Nobel Prize in Physics in 55 Years*: Donna Strickland was only the third woman to win the Nobel Prize in Physics, following Marie Curie (1903) and Maria Goeppert Mayer (1963). Strickland's award was particularly significant as it came after a long period without a female laureate in this category, highlighting her groundbreaking contributions to the field.

v. *Recognition of a Major Breakthrough*: The Nobel Prize in Physics was awarded to Strickland, Mourou, and Ashkin for their distinct but complementary advances in laser physics. While Strickland and Mourou were recognized for CPA, Ashkin was awarded for his work on optical tweezers, a technique that uses laser light to trap and manipulate biological molecules and cells.

Significance of Donna Strickland's Work

i. *Revolutionary Technology*: The CPA technique has revolutionized laser technology, enabling the development of extremely high-intensity lasers that are used in a wide range of applications.

ii. *Enhanced Precision*: Strickland's work has significantly improved the precision and capability of laser-based technologies, impacting medical procedures, manufacturing, and scientific research.

iii. *Inspiration for Women in Science*: Strickland's achievement has served as an inspiration to women in science and engineering, showcasing the critical contributions of women to cutting-edge research.

Term & definition

Chirped-pulse amplification (CPA)— a technique for creating ultrashort, yet extremely high-energy laser pulses necessary in a variety of applications—including the manufacturing of glass for cellphone screens.

Intensity - the measurable amount of a property, such as force, brightness, or a magnetic field.

Laser— light amplification by the stimulated emission of radiation.

Laser-Assisted in situ Keratomileusis (LASIK)— a laser-assisted surgical procedure for the correction of visual refractive errors.

Pulse— a single vibration or short burst of sound, electric current, light, or other wave.

Tissue— any of the distinct types of material of which animals or plants are made, consisting of specialized cells and their products.

4. Andrea Ghez – 2020

Andrea Ghez was awarded the Nobel Prize in Physics in 2020 along with Roger Penrose and Reinhard Genzel. The prize was awarded for discovering a supermassive compact object at the center of our galaxy. Andrea Ghez's contributions have significantly advanced the field of astrophysics, providing important insights into the nature of supermassive black holes and their role in the cosmos.

Key Highlights of Andrea Ghez's Nobel Prize

i. *Supermassive Black Hole Discovery*: Andrea Ghez, along with Reinhard Genzel, was recognized for her work on the supermassive black hole located at the center of the Milky Way galaxy, known as Sagittarius A*. Ghez's research provided critical evidence for the existence of this supermassive black hole and contributed to our understanding of its properties.

ii. *Research on the Galactic Center*: Ghez conducted detailed observational studies of the stars orbiting around the center of the Milky Way. Using advanced telescopes and adaptive optics, Ghez tracked the motion of these stars over many years. Ghez's observations showed that these stars were orbiting around an extremely massive and compact object, proving that this object was a supermassive black hole.

iii. *Adaptive Optics*: Ghez employed adaptive optics, a technology that corrects for the distortion caused by Earth's atmosphere, to achieve high-resolution images of the galactic center. This technology was crucial in Ghez's ability to observe the movement of stars close to the black hole and to infer its presence and mass.

iv. *Impact on Astrophysics*: Ghez's work has been instrumental in confirming the existence of supermassive black holes at the centers of galaxies. Ghez's research has advanced our understanding of how these objects form, their role in galaxy formation, and their influence on surrounding stars and matter.

v. *First Female to Win Nobel Physics in 55 Years*: Andrea Ghez was one of the two female laureates in Physics in 2020, along with Roger Penrose and Reinhard Genzel. Ghez's became the fourth woman to win the Nobel Prize in Physics since Marie Curie in 1903, highlighting Ghez's significant contributions to the field of astrophysics.

Significance of Andrea Ghez's Work

i. *Confirmation of Supermassive Black Holes*: Ghez's research provided key observational evidence for the existence of supermassive black holes, a critical component of our understanding of galaxy formation and dynamics.

ii. *Advancements in Observational Techniques*: Ghez's use of adaptive optics has revolutionized observational astronomy, allowing astronomers to study distant and faint objects with unprecedented clarity.

iii. *Influence on Galactic Studies*: Ghez's work has profound implications for understanding the structure and behavior of galaxies, including the Milky Way, and the role of black holes in the universe.

Term & definition

Adaptive optics (AO)— a technique of precisely deforming a mirror in order to compensate for light distortion.

Astronomy— the branch of science that deals with celestial objects, space, and the physical universe as a whole

Astrophysics— the branch of astronomy concerned with the physical nature of stars and other celestial bodies, and the application of the laws and theories of physics to the interpretation of astronomical observations.

Astrophysics— the branch of astronomy concerned with the physical nature of stars and other celestial bodies, and the application of the laws and theories of physics to the interpretation of astronomical observations.

Black hole— a region of space having a gravitational field so intense that no matter or radiation can escape.

Cosmos— The universe seen as a well-ordered whole.

Galaxy— a system of millions or billions of stars, together with gas and dust, held together by gravitational attraction.

Telescope— an optical instrument designed to make distant objects appear nearer, containing an arrangement of lenses, or of curved mirrors and lenses, by which rays of light are collected and focused and the resulting image magnified.

The Milky Way— the galaxy that includes the Solar System, with the name describing the galaxy's appearance from Earth a hazy band of light seen in the night sky formed from stars that cannot be individually distinguished by the naked eye.

Universe— all existing matter and space considered as a whole; the universe is believed to be at least 10 billion light years in diameter and contains a vast number of galaxies; it has been expanding since its creation in the Big Bang about 13 billion years ago.

5. Anne L'Huillier – 2023

© Nobel Prize Outreach

Anne L'Huillier was awarded the Nobel Prize in Physics in 2023 for her groundbreaking work in attosecond physics, a field that allows scientists to observe and manipulate the movements of electrons on an incredibly fast timescale. L'Huillier shared the prize with Pierre Agostini and Ferenc Krausz, together revolutionizing our ability to capture ultrafast processes in atoms and molecules. In addition to advancing our understanding of the quantum world, L'Huillier's work has paved the way for innovations that may transform various scientific and technological fields.

Key Highlights of Anne L'Huillier's Nobel Prize

i. *Awarded for Attosecond Physics Anne*: L'Huillier was recognized for her pioneering contributions to attosecond physics, the science of observing extremely fast processes at the atomic and subatomic levels. L'Huillier shared the Nobel Prize in Physics for developing techniques that generate attosecond pulses of light.

ii. *Groundbreaking Work on Attosecond Laser Pulses*: L'Huillier's research enabled the creation of laser pulses lasting only attoseconds (one billionth of a billionth of a second), allowing scientists to study the movement of electrons within atoms in real-time.

iii. *Leading the Way in Ultrafast Science*: L'Huillier's work is a cornerstone in the field of ultrafast science, helping to visualize and understand electron dynamics, chemical reactions, and fundamental quantum processes that occur on the shortest timescales ever measured.

iv. *Collaborative Nobel Prize*: L'Huillier shared the prize with Pierre Agostini and Ferenc Krausz, whose combined work has revolutionized our understanding of electron behavior, enabling real-time observations of quantum phenomena.

v. *Role as a Trailblazer for Women in Physics*: L'Huillier became one of the few women to win the Nobel Prize in Physics, continuing the legacy of female pioneers like Marie Curie and Donna Strickland.

Significance of Anne L'Huillier's Work

i. *Advancing Quantum Mechanics Understanding*: L'Huillier's development of attosecond pulses provides a crucial tool for exploring the quantum world, offering new insights into electron behavior and how atoms and molecules interact on the quantum scale.

ii. *Revolutionizing Atomic and Molecular Physics*: The ability to observe electron movements within atoms in real time has revolutionized the study of atomic and molecular physics, allowing researchers to witness and measure processes that were previously theoretical or unobservable.

iii. *Applications in Chemistry and Materials Science*: L'Huillier's work has potential applications in chemical reaction dynamics, enabling more detailed studies of how chemical bonds form and break. It also holds promise in materials science, where understanding electron behavior can lead to the development of new materials and technologies.

iv. *Impact on Future Technologies*: The attosecond technology developed by L'Huillier could have profound implications for quantum computing, ultrafast electronics, and advanced imaging technologies, pushing the boundaries of what's possible in modern technology.

v. *Shaping the Future of Time-Resolved Measurements*: L'Huillier's pioneering techniques have opened new avenues for time-resolved spectroscopy and experiments that can track ultrafast phenomena, providing tools for scientists to investigate the most fundamental processes in nature with unprecedented accuracy.

Term & definition

Attosecond— a unit of time in the International System of Units equal to 10^{-18} of a second.

Atomic— (of a substance) consisting of uncombined atoms rather than molecules.

Subatomic— smaller than or occurring within an atom.

Pulse— a single vibration or short burst of sound, electric current, light, or other wave.

Electron— a stable subatomic particle with a charge of negative electricity, found in all atoms and acting as the primary carrier of electricity in solids.

Quantum— a discrete quantity of energy proportional in magnitude to the frequency of the radiation it represents.

Quantum mechanics— the branch of mechanics that deals with the mathematical description of the motion and interaction of subatomic particles, incorporating the concepts of quantization of energy, wave-particle duality, the uncertainty principle, and the correspondence principle.

Chemical bond— the association of atoms or ions to form molecules, crystals, and other structures.

Material— the matter from which a thing is or can be made.

Quantum computing— computing that makes use of the quantum states of subatomic particles to store information.

Ultrafast electronics – a term relating to analog electronic systems associated with mm-wave and sub-mm wave generation and detection, signal conversion and basic device physics.

Imaging technology— the application of materials and methods to create, preserve, or duplicate images.

Time-Resolved Measurement— the measurement of dynamic processes in materials or chemical compounds by means of spectroscopic techniques.

Index

Adaptive optics (AO), 75
Agriculture, 30, 31, 39, 45, 47
Aluminum, 12
Antibiotic, 17, 24, 25
Astronomy, 75
Astrophysics, 74, 75, 76
Atomic, 25, 59, 64, 80, 82
Attosecond, 79, 82
Bacterial, 25
Biocatalyst, 32
Biochemical, 17
Biochemistry, 16, 18, 53
Biofuel, 32
Biological, 18
Biology, 25, 39, 51, 53
Bioorthogonal chemistry, 51, 53
Biotechnology, 32, 37
Black hole, 76

Cancer, 8, 38, 39, 47, 50, 53
Catalytic, 32
Cell, 26, 53
Cellular machinery, 26
Chemical bond, 83
Chemistry, 6, 8, 10, 11, 14, 17, 22, 25, 28, 29, 30, 31, 32, 35, 38, 43, 46, 48, 49, 50, 51, 59, 60, 81
Chirped-pulse amplification (CPA), 70
Click chemistry, 54
Compound, 18
Configuration, 65
Cosmos, 76
Cryogenic, 26
Crystallography, 15, 18, 23, 26
Degradation, 26
Deoxyribonucleic acid (DNA), 18, 40

Contributions of 55 Women Nobel Laureates
Part I - 12 Women Nobel Laureates in Chemistry and Physics

Diabetes, 18
Diagnostics, 52, 54
Diffraction, 19
Disease, 19, 54
Drug, 19, 54
Efficacy, 19
Electron, 65, 82
Enzyme, 33
Fertilizer, 33
Galaxy, 76
Genetic, 27, 41, 45
Glycobiology, 50, 54
Helium, 12
Imaging technology, 83
Immunology, 55
Industry, 30, 33
Infectious, 55
Insulin, 16, 19
Intensity, 71
Isotope, 12, 65
Laser, 67, 68, 71, 79
Laser-Assisted in situ Keratomileusis (LASIK), 68, 71
Material, 83
Medicine, 8, 16, 30, 37, 45, 55
Molecule, 19, 55
Neuroscience, 55
Neutron, 65
Nitrogen, 12

Noble gas, 65
Nuclear physics, 12, 66
Nuclear reaction, 66
Nucleosynthesis, 64, 66
Ore, 8
Organic solvents, 33
Penicillin, 15, 20
Pharmaceuticals, 30, 33
Pharmacology, 20
Phosphorus, 13
Physics, 9
Polonium, 7, 9, 58
Protein, 20, 31
Proton, 66
Pulse, 67, 71, 82
Quantum, 80, 82, 83
Quantum computing, 83
Quantum mechanics, 82
Radioactive, 9
Radioactivity, 7, 10, 58, 66
Radium, 7, 9, 58
Reaction, 33
Ribosome, 23, 27
Side effect, 56
Silicon, 13
Structure, 15, 20

Contributions of 55 Women Nobel Laureates
Part I - 12 Women Nobel Laureates in Chemistry and Physics

Subatomic, 82
Telescope, 76
The Milky Way, 76
Therapy, 20, 38, 46, 56
Tissue, 56, 71

Ultrafast electronics, 83
Universe, 77
Vitamin B12, 15, 20
X-ray, 14, 15, 17, 21, 23, 26

www.ingramcontent.com/pod-product-compliance
Lightning Source LLC
Chambersburg PA
CBHW070351230526
45471CB00006B/2512